WRITTEN BY
LAURENCE OTTENHEIMER-MAQUET, DIANE COSTA DE BEAUREGARD,
ALEXANDRE C. CZAJKOWSKI, ROGER DIEVART, MARIE FARRÉ, BEATRICE FONTANEL,
MARIE-PIERRE KLUT, CHRISTINE LAZIER, ANDRÉ LUCAS, PIERRE PFEFFER,
JEAN-PIERRE REYMOND, CHARLOTTE RUFFAULT, CATHERINE DE SAIRIGNÉ

COVER DESIGN BY
STEPHANIE BLUMENTHAL

TRANSLATED AND ADAPTED BY
BEVERLY A. HOBDEN AND ROBERT NEUMILLER

PUBLISHED BY CREATIVE EDUCATION
123 South Broad Street, Mankato, Minnesota 56001
Creative Education is an imprint of The Creative Company.

© 1990 by Editions Gallimard
English text © 2001 by Creative Education
International copyrights reserved in all countries.
No part of this book may be reproduced in any form without written permission from the publisher.

Library of Congress Cataloging-in-Publication Data
[Vie des animaux. English]
Amazing Creatures / by Laurence Ottenheimer-Maquet et al.
(Creative Discoveries)
Includes index.
Summary: Examines the characteristics and behavior of a multitude of animals,
from prehistoric sea creatures to the birds, mammals, and insects of today.
ISBN: 0-88682-945-3
1. Animals—Juvenile literature. 2. Animal behavior—Juvenile literature. [1. Animals—Miscellanea.]
I. Costa de Beauregard, Diane. II. Czajkowski, Alexandre C. III. Title. IV. Series.
QL49.0825   1999
590—dc21   98-7126

First Edition
2 4 6 8 9 7 5 3 1

AMAZING CREATURES

CONTENTS

PREHISTORIC ANIMALS	4
BIRTH AND BABIES	12
EVERY ANIMAL NEEDS A HOME	16
HOW ANIMALS EAT	24
CAMOUFLAGE AND COLOR	28
ANIMALS COMMUNICATE WITH EACH OTHER	36
NOCTURNAL ANIMALS	38
ANIMALS IN WINTER	40
MIGRATION	46
ENDANGERED ANIMALS	54
EXPLORE AND PLAY	61
GLOSSARY	70
INDEX	74

CREATIVE EDUCATION

The beginning of life on Earth

Many scientists believe the Earth was formed after a huge explosion in the universe. At first the Earth was very hot, and there was almost no oxygen in the air. This made life impossible. After millions of years, water began to collect on the surface and the seas formed. The first forms of life were tiny, one-celled creatures called bacteria.

Later, other small animals appeared in the sea. Most of these early animals are now extinct and remain only as fossils. Fossils of soft-bodied animals, such as jellyfish, sponges, and worms, are very rare. It's more common to find fossils of trilobites, snails, and other shelled creatures because of their hard coverings. Today, the nautilus is a relative of the primitive animals from prehistoric times.

Scientists say the Earth was formed about five billion years ago.

The first plants on Earth were algae, which evolved in the seas. Today there are about 280,000 different species of land and sea plants.

The first plants and animals lived in the sea.

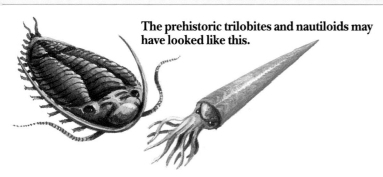

The prehistoric trilobites and nautiloids may have looked like this.

The first trees appeared hundreds of millions of years after small sea creatures appeared. By that time, many fish lived in fresh water rivers, lakes, and ponds. If the ponds began to dry up, the lack of oxygen in the water forced some fish to gulp air at the surface of the water and to develop lungs. These were the first amphibians, animals who live on land but begin their lives in the water.

Some animals began to lay their eggs on land. They were the first reptiles. Reptiles have scaly, waterproof skin.

Since reptiles are cold-blooded, they collect the energy they need from the sun. Many ancient reptiles are known today only through their fossils. However, most scientists believe all animals today are descended from this ancient stock. By changing and adapting to different climates and living conditions, new animals appeared while old ones became extinct. This is called evolution.

The long, skin-covered spikes on the dimetrodon's back collected or dispersed heat to help it maintain a constant temperature.

When the first amphibians walked on land, mosses and ferns were some of the most common plants on Earth.

The age of giants

There are many kinds of fossilized reptiles. Ichthyosaurs are one group of ancient prehistoric reptiles that spent all their lives in the sea. But they did have to come to the surface to breathe. Their bodies were streamlined and they had flippers, a back fin, and an upright tail fin. All these features made them excellent swimmers. Unlike other reptiles, which lay their eggs on land, ichthyosaurs gave birth to live babies in the water.

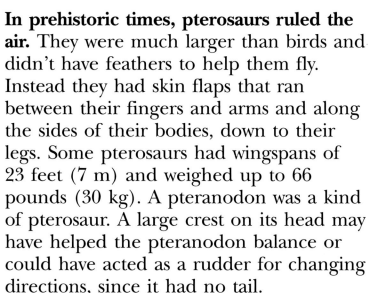
Pteranodon

Ichthyosaurs resembled today's dolphins.

In prehistoric times, pterosaurs ruled the air. They were much larger than birds and didn't have feathers to help them fly. Instead they had skin flaps that ran between their fingers and arms and along the sides of their bodies, down to their legs. Some pterosaurs had wingspans of 23 feet (7 m) and weighed up to 66 pounds (30 kg). A pteranodon was a kind of pterosaur. A large crest on its head may have helped the pteranodon balance or could have acted as a rudder for changing directions, since it had no tail.

Plesiosaurs had long necks and small heads. Their jaws were full of sharp, pointed teeth. They probably caught and ate fish. Some people think the Loch Ness monster in Scotland is a plesiosaur.

A mosasaur attacking plesiosaurs

The amazing dinosaurs

Dinosaurs are among the strangest animals that have ever lived. When people first found dinosaur bones, they thought the bones belonged to giants. It wasn't until the 19th century that Sir Richard Owen and other scientists realized the bones belonged to reptiles. These ancient reptiles were named dinosaurs, which means "monstrous lizard" in Greek.

Many different kinds of dinosaurs once lived, and their remains have been found all over the world. Brachiosaurus was one of the largest. It weighed 80 to 85 tons (66–77 t) and was 40 feet (13 m) tall—as tall as a four-story building. But not all dinosaurs were giants. Compsognathus was a fierce predator, even though it was only the size of a dog.

Because of its long neck, Diplodocus was able to feed on leaves from the tops of trees.

They ruled the Earth for 160 million years . . .

Dinosaurs laid eggs. Some fossilized dinosaur eggs have been found. Proceratops, a plant-eating dinosaur, laid her eggs in a shallow nest she dug in the ground.

Then she left them, just as today's turtles do. Maiasaurus, on the other hand, stayed with her eggs as crocodiles do today. She may have brought food back to the nest for her young after they were born.

The shapes and sizes of dinosaurs varied widely. Many of them, especially the plant eaters, had bony plates embedded in their skin. This armor helped to protect them when they were attacked.

Stegosaurus had a row of bony plates along its spine for protection and to gather heat from the sun. The skin covering these plates was filled with blood vessels. The sun warmed the skin, and the blood carried the heat from the skin through the stegosaurus's body.

Slow-moving brontosaurus (1), stegosaurus (2), and the ornithopods (3) all ate plants. Meat-eating allosaurus (4) hunted for prey.

...and then disappeared.

Some of the plant-eating dinosaurs, such as triceratops, had sharp horns on their heads. Triceratops resembled a rhinoceros and may have charged at its enemies in much the same way. Stegosaurus and the ankylosaurids had enormous spikes at the end of their tails, which they could swing against attackers. This tough armor was a defense against the long sharp teeth and slashing jaws of the meat-eating dinosaurs.

Why dinosaurs disappeared is one of the great mysteries of the world. Some say that the Earth must have been shaken by terrible volcanic eruptions. According to this theory, poisonous gases from the eruptions killed all the plants. Without plants to feed on and give off oxygen into the atmosphere, the animals would have died, too. But this can't be the whole answer, because some animals did survive.

This ankylosaurid was well-protected.

Another theory is that a meteorite collided with the Earth, causing a massive dust cloud that blocked the sunlight. Earth's temperature dropped dramatically. Cold-blooded dinosaurs couldn't survive the lower temperatures, but some of their relatives, like crocodiles, snakes, lizards, and tortoises, did. So this theory doesn't provide the entire answer either.

The mammals did survive. Unlike reptiles, mammals produce their own heat and can keep their bodies at a steady temperature. A covering of hair, fur, or wool insulates them from the temperature of the air around them. They could have lived easily through the cold. After dinosaurs died out, mammals became bolder and bigger and adapted to the changing conditions. They replaced the dinosaurs as rulers of the Earth.

A march of the dinosaurs:
1. Anchisaurus 2. Mamenchisaurus
3. Stegosaurus 4. Allosaurus 5. Deinonychus
6. Ankylosaurid 7. Spinosaurus
8. Edmontosaurus 9. Chasmosaurus

Early mammals

Fossilized fish and frog that lived at the same time as the first mammals

The first mammals appeared 220 million years ago. They evolved from reptiles, adapting to their changing surroundings. As dinosaurs were disappearing 65 million years ago, mammals were increasing.

Paleontologists find fossilized bones.

The first mammals were small, most only about the size of a present-day mouse. They hid by day to keep out the way of predators and fed by night on slugs, worms, and other small animals. All mammals have teeth of different sizes and shapes for chewing, biting, and tearing. The roofs of their mouths are made in such a way that they can breathe and eat at the same time. They are covered with fur and have mammary glands that produce milk to feed their young.

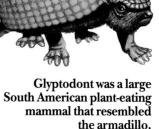

Glyptodont was a large South American plant-eating mammal that resembled the armadillo.

Between one and two million years ago, great ice ages began to grip the Earth. In these icy conditions large mammals, such as the saber-toothed tiger, flourished. The huge, curved, stabbing teeth of this cat were eight inches (20 cm) long. The saber-toothed tiger is now extinct.

Vinatherium was an early mammal that looked a little like a rhinoceros. It had three pairs of horns on its head.

Some extinct mammals have descendants living today.

Mammoths are related to the elephant, but stood more than 14 feet (4.5 m) high and had curving tusks up to 13 feet (4 m) long. Today elephants are the largest living land animals, but the biggest range in size from only 10 to 13 feet (3–4 m). Mammoths whose fossil remains have been found still capture people's imagination.

Mammoths have been found frozen in the ice, perfectly preserved, in Siberia and other areas of the far north. Until these recent discoveries, no one had seen mammoths since early humans had hunted them for food. On the frozen mammoths, scientists examined not only the skin, but also the heart, brain, and other internal organs that normally decompose. Scientists could even look into the mammoth's stomach and see what it had eaten for its last meal.

Changing with the times

While mammals evolved, so did other animals. Small differences from one generation to the next help animals adjust to changes occurring in the world around them. Different skills, behaviors, or variations in appearance shown by some individual animals enable them to survive, reproduce, and eventually pass these traits on to the next generation. Without this ability to adapt to new conditions, animals and plants die out, or become extinct.

Male and female animals of the same species must mate in order for offspring, or babies, to be produced. The male's sperm fertilizes the female's egg, and a new life begins. How the male and female meet, whether they stay together as a family, or if they part immediately varies tremendously in the animal kingdom. The ways parents look after their young, if they do so at all, and how long the young need to be cared for also differs from one species of animal to another.

Mothers have many different ways of caring for their offspring. 1. Elephant 2. Kangaroo 3. Crocodile 4. Koala 5. Monkey 6. Lion 7. Beaver 8. Grebe 9. Swan 10. Scaly anteater

Animal families can be either large or small.

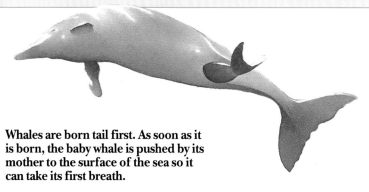

Whales are born tail first. As soon as it is born, the baby whale is pushed by its mother to the surface of the sea so it can take its first breath.

Female fish and amphibians may release thousands of very small eggs into the water. There, the eggs are fertilized by the male's sperm and develop very quickly—fish eggs into fry and frog eggs into tadpoles. Many eggs must be produced because most of these tiny animals are eaten before they grow up.

In many other animals, the male fertilizes the eggs inside the female's body. In the case of birds, reptiles, and some rays and dogfish, the fertilized egg grows a tough shell. The egg is then laid in a safe place, where it later hatches.

Sometimes the fertilized egg stays inside the mother until the baby is ready to be born. Salamanders, skinks, and some snakes do this. Most mammals, like humans, have a placenta that uses an umbilical cord to join the blood supply of the mother with that of the baby.

Some mammals, like kangaroos and opossums, are called marsupials. At birth, marsupial babies are tiny and not fully developed. After they're born, the young crawl inside a pouch on their mother's abdomen. They stay inside the pouch and feed on their mother's milk until they have grown enough to survive outside.

Why aren't all offspring born looking like adults? If the egg the mother produced is very small, there isn't enough food in it for the offspring to develop much before hatching. Some eggs, like those of most insects, hatch into larvae. This is a stage for eating. Caterpillars are moth or butterfly larvae; maggots are fly larvae. The change that they go through is called metamorphosis.

Not all parents stay with their young.

If a species is to survive, each adult must have at least one offspring. Since not all animals are able to look after their young, they produce enough offspring to make sure that some will survive to adulthood. A female cod lays a million eggs in the open sea and never sees its young. Only a few may survive. Mammals and birds, on the other hand, care for their young after birth. The young have a better chance for survival.

Bears usually give birth to twins. Young bears stay with their mother for up to two years.

Each species has its own method of raising offspring. After the female midwife toad lays her eggs, the male attaches them to his body. When the eggs are ready to hatch, he takes them to the water and the newborn tadpoles swim away. Gorillas and chimps typically look after their young for four or five years. In most cases, human offspring have contact with their parents all of their lives, but often begin providing for themselves at age 18.

Female crocodiles protect their young. When explorers first saw young crocodiles peeping out from the adult's mouth, they thought the youngsters were being eaten. Actually, the young enter their mother's mouth for protection when danger threatens. Despite such parental care, only a few young crocodiles survive. But those that do survive may live for 50 years and have several offspring themselves.

Suckling lion cubs may feed from their own mother as well as from other lionesses in their pride that have cubs.

Mammals and birds are caring parents.

Baby elephants typically take shelter beneath their mothers when danger threatens.

Who looks after the young? Warm-blooded creatures, mammals and birds, devote more time than reptiles do to raising their babies. Young mammals need their mother's milk, and baby birds need the food their parents bring. Baby birds and mammals depend on their parents for a long time before they can face the world on their own.

The mother elephant will chew leaves and then put some in the youngster's mouth. By sharing, young animals learn which foods are good to eat.

Animals that live in groups can share the task of looking after the young. Lionesses in the same pride are usually related, and some will "baby-sit" while the other mothers are out hunting. This extra care gives the cubs a better chance for survival.

Young tigers learn to fight by playing with each other and with their mother.

With many bird species, both parents feed the young. But once the youngsters have learned to fly and leave the nest, they may never see their parents again. Young birds must rely on instinct to provide them with the knowledge not learned from their parents. They generally don't have to learn how to build a nest, for example. When the time comes to start their own family, they just seem to know what to do.

When two pairs of ostriches meet, they will sometimes fight for the guardianship of each other's chicks.

Shelter from the outside world

Just as humans do, animals need protection from the weather. Many have this protection ready-made, either in the form of a shell that they carry on their backs or a thick outer coat. Even so, they may still seek out or build a sheltered spot to rest or sleep in, to give birth in, and to die in. Others, like humans, must always rely on shelters they build to protect them.

Beavers build sturdy lodges to house their families.

Rabbits and marmots dig burrows, bears make dens, squirrels make nests, and beavers build lodges. Birds build nests, but only to house their eggs. Snakes spend the winter under stones or in caves. Frogs and toads hide in gardens or ponds and bury themselves underground to survive extreme temperatures. Some animals don't have to build anything because they're born with a home on their backs. When a snail hatches from its egg, it already has a tiny, transparent shell on its back. Oysters and mussels have two-part shells protecting their bodies.

Some animals live together, sharing their home. In an anthill, the larvae are kept warm in a special chamber. After a few weeks they change, or metamorphose, into worker ants and begin to enlarge the anthill, using twigs, moss, and their own saliva. They make corridors and even gates that they close at night.

In hives, bees build honeycombs from wax. Worker bees keep the hive clean, removing dirt and any bees that have died during the night. Some bees guard the hive, while others go out to gather food.

Ants work to improve their anthills.

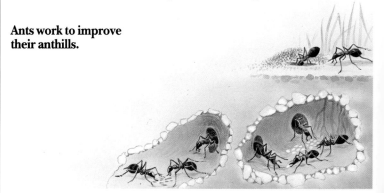

Homes can be many shapes and sizes.

Some animals camouflage their homes so well that it's hard to see them. A squirrel's nest high in the branches of a tree looks like a little bowl woven from sticks and leaves. In Africa, chimpanzees make sleeping platforms by weaving leafy branches together. They use their platforms for only one night.

The spider makes its silken web by spinning thread from special glands called spinnerets on the rear of its body. The first threads form a Y, around which the rest of the threads are connected. When the web is complete, the spider waits, ready to strike when an unsuspecting insect blunders into the web. The silk produced by spiders is the strongest natural fiber known.

Some fish make nests. The male stickleback makes a nest tunnel. When the tunnel is complete, he attracts a female and brings her to see it. If she is impressed, she will lay her eggs in it.

A chimpanzee sleeps on its platform of branches.

A spider inspects its web for damage.

Sea otters sleep wrapped in seaweed.

Birds are master builders.

On the right, the male bowerbird builds elaborate nests to attract females. Below is a nest the titmouse might make.

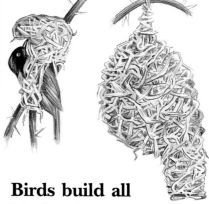

Birds build all kinds of nests.
They build nests only when it is time for them to breed. The nest makes a safe cradle for the eggs and chicks. Once the young birds have begun to fly, or are fledged, the nest will be abandoned. Swifts return to the same nest each year, but blackbirds always build new ones. The cuckoo doesn't build at all, but lays her eggs in another bird's nest.

Nests can be a hollow in the ground or a basket woven from twigs; they can be perched on a cliff or laid down among pebbles on the beach. Each species of bird has developed a building style best suited to it.

There are many ways to build a nest.

An eagle on its aerie

Weaverbirds know how to make knots and loops. They use grass and other types of vegetation to weave their nests.

The golden eagle and other birds of prey make enormous nests on cliff faces. They build these nests out of large sticks and branches that they carry in their talons. These nests, or aeries, can be 10 feet (3 m) wide.

The Asian long-tailed tailorbird makes holes with its beak along the edges of a large leaf, then stitches it into a bag using grass for thread. It fills the bag with padding to protect the eggs.

Barn swallows return from their wintering areas and use balls of clay, straw, and blades of grass to build their nests in caves or buildings in the spring. They fill the nest with feathers, down, and hair to keep their chicks warm.

Like many water birds, this giant South American coot builds a floating nest just as grebes do.

Woodpeckers use their sharp, chisel-like bills for drilling into the bark of trees to hollow out a nest. They have special gripping toes and stiff tails to support themselves while they work. Their beaks and skulls are specially adapted to absorb the shock of drilling.

Barn swallows bring hay and straw to line their nests.

Some mammals make burrows...

Many mammals tunnel underground to find shelter. There they are safe from attack by other animals and find protection from the weather. However, they usually must leave their burrows to find food.

Badgers live in burrows that they dig in the forests or on the prairies. Badgers have large heavy bodies that look ill-suited to do this kind of work. But they are swift diggers; their burrows are a complex series of tunnels having one main entrance and emergency exits for use only when danger is near. The entrance leads to several inner chambers, which are well-ventilated. A badger's burrow can be as much as 30 feet (9 m) deep.

Badgers are born underground and spend the first few weeks of their lives in the safety of the burrow.

. . . that they line with grass and moss.

Weasels often use the tunnels of shrews, rabbits, and other animals as makeshift homes.

Deep in the rabbit burrow, the doe gives birth. She has already prepared a cozy nest for her young, pulling fur from her own chest to line it. This fur will keep the babies warm when she has to leave them to find food. When she leaves, she carefully closes the entrance to the burrow. She will return to her young only to feed them. After three weeks, they are ready to feed with her in the fields.

Moles live alone in a maze of tunnels, and they will seek out other moles only during mating season. They use their strong, shovel-like paws to dig and catch worms, relying only on smell and touch and vibrations picked up by their whiskers. Moles eat many small animals, but earthworms are their favorite.

The duck-billed platypus, an egg-laying mammal, lives in burrows along riverbanks.

Many other small mammals live in burrows that protect them against extreme weather conditions. Under the desert surface, the sand is much cooler. Fennec foxes of northern Africa dig burrows to protect themselves from the scorching sun. They are nocturnal animals. They spend their days resting underground and only come out at night to feed.

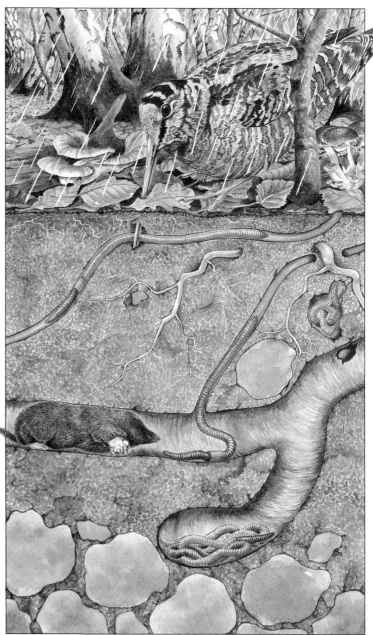

The beaver is a skilled architect.

Beavers use their front teeth for gnawing wood. These teeth never stop growing, but they are worn down from hard work.

Beavers work through the night, cutting trees by the water's edge. Their strong, curved, orange-colored incisor teeth can gnaw through branches and small trees. Using sticks and branches held together with mud and stones, beavers make dams across streams.

Beavers build resting places and eating platforms all around their territory. Underwater canals allow them to pass from one pool to another without ever touching the ground.

Hard-working woodcutters, beavers cut down many trees before their task is finished. A beaver can cut down a five-inch-thick (12.5 cm) willow tree in three minutes. Standing on their hind legs, they gnaw around the trunks, making a shape like a sharpened pencil. When the tree falls, the beavers cut it into pieces they can easily move. Branches are used for building; the leaves, bark, and small sticks are either eaten immediately or stored underwater to be eaten during the winter.

How does a beaver build its dam? The beaver wedges large branches into the riverbed against the current. Smaller branches are placed on top, and the gaps are filled with mud, stones, and small plants. The beaver constantly checks the dam's walls and repairs them immediately if they break. Behind the dam, the river widens into a safe pool. Here beavers build their lodges using the same materials as they used on the dam.

The beaver's chisel-like teeth gnaw easily through a tree trunk.

An underwater entrance leads to a dry nest.

Beavers communicate with each other

The beaver's living quarters, or lodges, must be strengthened before winter. Small rocks and mud are mixed into a cement to fill any cracks that have appeared during the summer. The beaver smooths out this mixture to give it an extra waterproof coating.

Inside the lodge, everything is neatly arranged. The entrance is always underwater, so the beaver must shake itself dry as soon as it gets inside. Then it goes from the hallway into the main room, where the dry nest is. This room is always warm, even when the outside temperature is freezing. A vent in the ceiling allows air to circulate so that the air inside the lodge stays fresh. Beavers spend most of the winter inside the cozy lodge. They leave only for food.

Beavers change the bedding in their lodges often.

The mother carries her kit on her back or her tail.

She can also swim while holding her kit in her mouth.

Beaver parents stay together for life. Each year, the mother has a litter of up to five kits. Beavers know how to swim soon after they are born, even though they rarely leave the lodge in the first few weeks. The kits stay with their parents for up to two or three years. Older kits help look after their younger brothers and sisters and help gather food for winter. When they have learned all their parents can teach them, the kits leave to find mates and set up their own lodges.

An island lodge made of branches and mud

Teeth are used for killing and eating.

All animals need to eat so that they stay healthy and strong enough to survive all the difficulties that face them. Some animals search for plants and fruit, while some hunt other animals. Teeth are special tools to help animals eat. By looking carefully at an animal's teeth, we can tell what kind of food it eats and the type of life it leads.

A shark's teeth are razor sharp.

Hyena   Jackal   Tiger

Animals that eat meat are called carnivores. They have long, pointed front teeth called canines that they use when catching and killing prey. The pictures above show these teeth, like sharp daggers, at the front of the animals' mouths. When they want to eat, carnivores have to turn their heads sideways and use their sharp cheek teeth to cut meat off a carcass.

Insectivores are meat eaters that eat small animals such as beetles, worms, and butterflies. Shrews, hedgehogs, moles, and most bats are insectivores. They have canine teeth for stabbing and sharp cheek teeth for slicing.

Animals that eat plants are called herbivores. Some herbivores graze on grass, while others browse on bushes and trees. Both have cheek teeth called molars, with flat surfaces that grind and squash plants as they eat. Some animals called omnivores have teeth that allow them to eat both plants and animals.

A rabbit's front teeth grow all its life and are worn down by eating. This is what the teeth would look like if they didn't wear down.

Animals have different kinds of teeth.

Raccoons, rats, guinea pigs, and porcupines all have teeth specially adapted for their way of life.

Rodents have chisel-like incisor teeth that grow constantly. Some of them, like hamsters, can carry food in their cheek pouches. Rodents eat many different kinds of food and are found all over the world.

Some plant eaters chew cud. Plants aren't easy to digest, so some herbivores, such as cows, sheep, and deer, eat in two stages. First, they crop the plants, chew them, and swallow them. Later when they are resting, small amounts of food, called cud, are regurgitated back from their stomachs into their mouths, and the animals chew the food again. The animal gets more nutrition from the food when it chews the food repeatedly.

An African elephant's molar has diamond-shaped ridges. The tooth is one foot (30 cm) long and three inches (7 cm) wide.

Some fish have many sharp teeth. In most cases, the teeth curve inward toward the fishes' throats to stop their prey from quickly escaping.

Most sharks have sharp, triangular teeth that grow in rows around their mouths. Small, new teeth grow on the inside row and get bigger as they move toward the edge of the mouth. These new teeth replace those that are lost.

Each type of animal has teeth that suits its diet. Sharks that bite their prey have serrated edges on their teeth, while those that swallow their prey whole have sharply pointed teeth for grabbing and holding. Some sharks and rays eat only shellfish and sea urchins. They need flat teeth to crush their prey.

Horses use their tongues and lips to select food. When they chew, their jaws move from side to side. They don't chew cud. Their teeth are adapted for grinding coarse grass.

Tusks and fangs are teeth, too.

Some animals have canine teeth that have grown into long tusks. Tusks are often curved and are usually very strong. The sight of them is often enough to scare off predators. Males may also use them to show off to females. Elephants, walruses, and some types of pigs, such as wild boars, have tusks. Walruses anchor themselves with their tusks when climbing on ice.

Walruses use their tusks to scrape barnacles and shellfish from rocks.

Elephants are often killed illegally for their tusks. Some people admire this ivory and want to own objects made of it. Elephants need their tusks not just for protection, but also to dig for water during droughts and to pry bark from tree trunks for food.

Snakes have fangs. All snakes are carnivores. Boa constrictors and pythons, both snakes with large bodies, kill their prey by curling around them and squeezing the breath out of them. Thin-bodied snakes, such as rattlesnakes and adders, are often poisonous. They strike at their prey, biting with hollow fangs that deliver a dose of poison.

Snakes swallow their prey without chewing. A snake's jaw comes apart at the front and at the side where it joins the skull. This allows the snake to open its mouth wide to swallow its prey, such as mice and other rodents, whole. It then has to rest a long time to digest its food.

A 1990 international treaty banned the sale of ivory worldwide.

Pit vipers have curved fangs.

Some animals have no teeth.

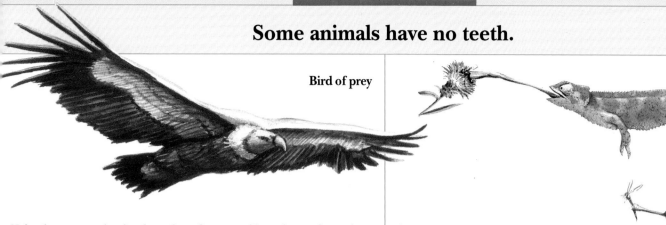

Bird of prey

Chameleon

Birds use their beaks for collecting food. Hooked beaks, like those of eagles, hawks, and other birds of prey, are for tearing. The slender beaks of the woodpecker or hummingbird are for stabbing and probing. Stubby, triangular beaks of finches and sparrows are for cracking seeds. The various birds in a backyard or park each have different types of beaks depending on the food they eat.

Some animals use their tongues to catch their prey. A chameleon has a very long tongue, which shoots out of its mouth to catch insects. Frogs and toads have shorter, wider tongues that also work well for catching insects. Their tongues are connected at the front of their mouths and flip out at lightning speed. Snails have a tongue-like organ called a radula. The radula contains small rows of teeth which act like a cheese grater to scrape away at leaves and flowers. Some snails can destroy plants quickly, so many gardeners don't like them.

Sea creatures get their meals in many ways. Crabs and lobsters use their pincers to grab prey and break it open. The seahorse has a mouth like a vacuum cleaner. It uses its tail to attach itself to seaweed and sucks up any food that passes by. Some whales, such as the blue whale and baleen whale, have no teeth. Instead, they have plates of a horny material called baleen hanging down inside their mouths. The edges of the baleen are fringed. When these whales are hungry, they draw large amounts of water through the baleen and strain out tiny sea animals called plankton.

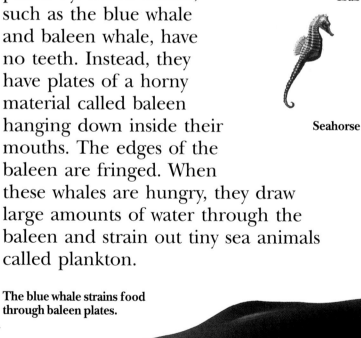

Frog

Snail

Crab

Seahorse

The anteater has no teeth. It uses its sticky tongue to lick up ants and termites.

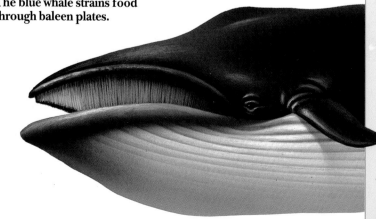

The blue whale strains food through baleen plates.

Many animals can disguise themselves.

The fawn's dappled coat hides it in the undergrowth where it's safest from danger. Many animals depend on their coats for camouflage. Some animals are camouflaged to help them hunt without being spotted, and others have camouflage to prevent them from being seen by predators.

The adult honey badger and the cheetah cub have a similar type of camouflage: fur on their backs blends in with their surroundings.

Young animals are often weak and helpless. Sometimes they are left alone while their mothers find food. So it's important that they hide well. This is safer for some animals than trying to run from danger. The silvery fur along a cheetah cub's neck and back helps it blend in with the tall, waving grass.

Color and shape help animals blend in with their backgrounds. The glassy-wing skipper butterfly is almost invisible in the picture at left. Blending into the background gives protection only if the animal stays still. The sulfur butterfly is easy to see once it moves. Other animals are hiding in the picture. Look carefully for the frog with eye spots on its legs, the moth that blends in with the moss on the tree trunk, and the bird sitting on the branch.

Sulfur butterfly

Some animals match their backgrounds.

To remain disguised, some animals change color with the seasons. The arctic fox is brown in summer to match the rocks and vegetation, but white in winter to match the snow. Many arctic animals change in the same way. The coloring of snowy owls, on the other hand, isn't so straightforward. They stay the same color year round, so their feathers are a combination of summer and winter color—a dappled white and brown.

A suit of greens and browns hides the chameleon.

Chameleons can change color as they move. Some types of flatfish, such as the flounder, can do the same.

1. Arctic fox 2. Rock ptarmigan 3. Weasel 4. Arctic hare 5. Snowy owl

Animals' spots and stripes hide their shapes.

Camouflage is important for hunter and hunted alike. It can range from simple markings to elaborate disguises. A zebra's stripes change direction over its body. This may make zebras more difficult to see at a distance for the animals who prey upon them. Stripes and dark patches in unexpected places make an animal's body harder to recognize as a whole.

These camouflaged animals live in Africa.

Predators must stalk their prey without being seen, using bushes and leaves to give them cover. The leopard resting in the tree may be hard to see in the picture below. Lionesses hunt in the open. Their sandy brown fur blends with the tall grasses.

Lone hunters need a good disguise if they want to catch their prey. Tigers hunt alone in forests and along reed-covered riverbanks. Their black stripes help them blend into the sunlight and shadow as they wait to pounce on animals that come to drink.

The tiger's face has broken stripes.

It is summer year round in tropical forests. It's hot and wet all the time. Plants flower and fruit throughout the year and never lose all their leaves. This makes a rich home for animals.

Many wingless insects and snakes escape detection by matching the green-brown background of the trees. On the forest floor, spots and stripes disguise both prey and predators. However, many birds are brightly colored. They fly away from danger rather than hide from it.

The jerboa's sandy back matches the color of its desert home.

All disguises help animals survive.

The more closely an animal matches its background the greater its chances of survival. Desert animals, like the African antelope above, are often sandy in color with a bold stripe on their sides. From a distance it is hard to tell that they are animals at all.

Jerboas and fennec foxes are nocturnal, which means they are active at night. Their beige color helps them match the colors of the desert at dawn and dusk. Scorpions and lizards may be yellow and white in sandy deserts, but their cousins living on the rocky outcrops are black or gray. Even cheetahs that hunt along the desert edge tend to be less spotted to blend in better with their surroundings.

The three-toed sloth of Central and South America spends most of its life hanging upside down in trees. Millions of tiny green algae grow in its fur, making it nearly invisible among the branches. The pronghorn antelope of the western United States is nearly odorless for the first few weeks of its life, making it hard for predators to find.

Pretending to be dead is another way to go unnoticed. Because dead animals may be diseased, most predators will strike only at live prey. Many frogs, snakes, young birds, and mammals pretend to be dead when a predator approaches. They let their bodies go limp and stay absolutely still fooling their predators. This works well for young, injured, and defenseless animals.

Fennec foxes live in the Sahara Desert, but come out only at night to feed. The color of their fur makes them well suited for nocturnal life.

Trickery on the sea bed

It's extremely important for animals to be well hidden on the sea bed. The ghost crab hidden among the coral in the picture becomes visible to its prey only when it moves. Its transparent color helps it blend into the sea floor. The crab spends most of its time sitting very still, waiting for food to come its way.

Some animals make their own disguises. They gather fragments of shell or seaweed and even let other animals grow on them. The hermit crab may live inside the shell of a dead snail, or a sea anemone may live on a crab's back. Sea animals have many disguises to keep them safe from predators.

A sponge is growing on this crab's back.

Seeing isn't always believing.

The stonefish deserves its name. When half buried among rocks, it's almost impossible to see. That's not its only protection. The spines on its dorsal fin also eject a powerful poison.

   A fish's location often determines its camouflage. Ocean fish have dark backs and lighter bellies that make them difficult to see from below. However, tropical, shallow-water fish have many curious disguises. Some have spots by their tails that resemble eyes.

An octopus can change its color to blend with its background. First it looks like a rock, but in a moment it can match the color of the coral that it passes. Its prey must be careful; the octopus is easy to miss because it blends into its environment so well.

What are these straight shapes below? While they look like seaweed waving in the tide, the razor fish are swimming head down to avoid being recognized.

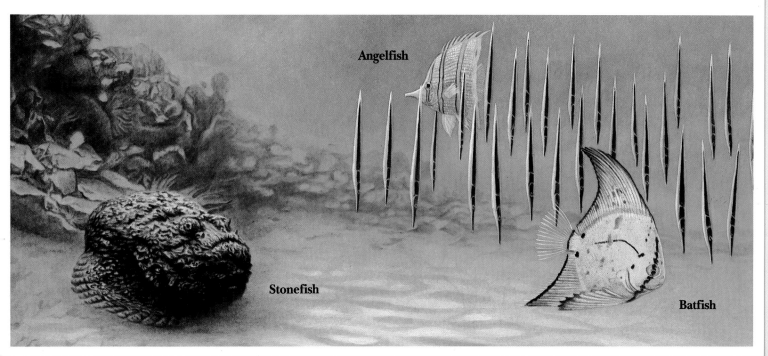

Angelfish

Stonefish

Batfish

Colors are signals . . .

Many animals use colors to communicate. Colors can be like a uniform, showing that an animal belongs within a certain group. Colors can also show who's in charge. They can be warnings, too, as well as aids to courtship. The brightly colored face and buttocks of this male mandrill, for example, indicates that he is the leader of his group.

Mandrill

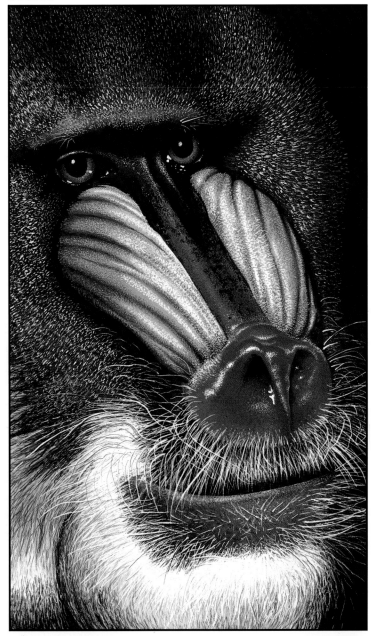

A male frigate bird (right) is displaying to his partner.

Red is often a favorite color for many birds. During mating season, the male frigate bird shows off his brilliant red throat pouch to attract a mate. Birds with young will give up their food when they see the gaping red throats of their chicks. Adult seagulls have a red spot on the underside of their beaks. Chicks are attracted to this spot and peck it. This triggers the parents to give up food they are carrying in their beaks.

Female birds are often less brightly colored than males. Their dull colors hide them when they're looking after their eggs or chicks.

In one of the most magnificent displays of color, the peacock spreads his tail to impress the peahen.

... for mating and warning of danger.

Generally, only animals that are brightly colored can see in color. Others just see black, white, and shades of gray. Nocturnal animals such as badgers see in black and white. Colors are hard to distinguish in the limited light of nighttime, so it was unnecessary for many animals to develop the ability to see their world in color.

The ring-tailed lemur uses his striped tail to signal to others in the group.

At night, fireflies give off a green-yellow to red-orange light. Only the adult male fireflies can fly. The adult female firefly and larvae are called glowworms. To attract a mate, fireflies signal with their glowing abdomens.

Light is also given off by some fish that live in the deepest parts of the ocean, where it's totally dark. These fish with luminous particles on their bodies often use their light to attract mates and prey.

The bowerbird collects fresh flower petals to attract his mate.

Some colors warn of danger. Red, yellow, and black usually warn of danger in nature. Wasps and bees have black and yellow stripes that advertise their presence and warn off enemies. Harmless animals can use these danger colors as protection. If a small beetle is red or yellow and black, predators will avoid it, in case it is poisonous or stings. Predators have learned through experience that these colors mean something unpleasant.

Fireflies

Animals communicate in many different ways.

Information can be given by sound, scent, or facial expressions and gestures. Solitary animals mark their territories to tell others to keep out. Animals that live in groups have distinctive scents that allow them to be identified by other members of the group when they come home.

Sounds often are used to greet and reassure, or to frighten and put group members in their place. Most monkeys and apes live in groups and have developed several types of "language" to help them communicate with each other.

Wolves howl to keep in touch with their pack as they travel through the night. The sound tells other wolves how large the pack is and where its territory is. Each wolf has a rank in the pack. When greeting a higher-ranking male member of the pack, a male wolf will lower his tail and lay his ears back.

Male gorillas beat their chests and make loud growls to frighten off enemies. They look very frightening, but they charge only as a last resort when their enemy ignores their warning.

Grooming, other forms of touch, and facial expressions are used to form bonds between individual animals. Animals also play together to show affection.

Monkeys and apes express their feelings in ways similar to human beings. Humans also show a great deal of emotion with their faces. If people are happy, angry, or hurt, their faces often show it, and friends may "read" the expression and try to help.

Chimpanzees use facial expressions to show how they feel.

Some animals use scent and sound to indicate food or danger.

These ants are following a scent trail laid over a leaf by a member of their group. Turn the leaf, and the ants turn too. They keep following the trail.

Ants leave each other scent messages. By laying down a scented trail, which shows the others the direction to travel, an ant informs other members of the anthill where food can be found. Hundreds of ants will then follow the trail and help bring the meal home. By working together, ants can achieve much more than they could alone. If one ant is attacked, for example, hundreds will come to its defense.

Many animals leave scent messages to mark their territory and warn others away. Hedgehogs live alone and like it that way. Moles don't welcome strangers into their burrows, and shrews fight fiercely when they meet. Skunks use a strong scent as a means of defense. The skunk's scent is so strong that it can blind its enemies. Reading the signs others have left helps animals avoid trouble and prevents them from injuring each other.

Hedgehogs mark their territories to keep other hedgehogs away.

Some rabbits live in groups called warrens. In each warren, one rabbit acts as a lookout. With ears pricked, the rabbit is alert to any danger. To warn the others, it thumps the ground with its hind leg before running away. Its bobbing white tail is a further signal for others in the warren to seek the safety of their burrows.

Nighttime adventurers...

As night falls, most animals get ready to sleep. But others are just waking up from a good day's rest. Porcupines, foxes, badgers, and owls are just some of the animals that love the dark. To hunt at night, these nocturnal animals have developed very keen senses of hearing, smell, or sight.

An owl's eye: in darkness, in twilight, and at daybreak

Nocturnal animals sleep during the day. Some, like owls, have large eyes, and therefore need little light to see. In fact, if you were to shine a flashlight in an owl's eyes, the black pupils in the center would contract to become as small as a pinhead to block out almost all the light. Its eyes are too sensitive to allow in such a massive dose of light. Even daylight is too strong for owls to feel comfortable.

Some nocturnal animals rely very little on their eyes. In the darkness of night, animals' other senses are sometimes better developed and more useful. Bears, for example, have a keen sense of smell which they use to find their food.

In the still of the night, the slightest noise can be a warning. Special feathers at the edge of owls' wings allow these birds to fly silently. Due to these feathers, the owls' prey won't hear them approaching with their talons outstretched. After the owl has killed a mouse, it returns to its feeding post, carrying the dead mouse in its beak.

During the day, owls rest quietly in trees or even old buildings, sheltered from the light. At night they are wide awake and ready to hunt for mice and other small rodents.

... have senses that can penetrate the darkness.

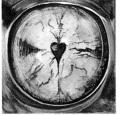 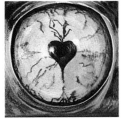    

Frogs and toads see best at night when their pupils open wide.

The pupils of a nocturnal snake close in daylight and open at night.

Insect-eating bats catch their prey by using echolocation. They make as many as 60 high-pitched squeaks per second while they fly, searching for prey.

The nocturnal hawk moth flies silently and has very sensitive hearing.

The bat sends out sounds, which are too high for humans to hear, in front of where it is flying. If an insect crosses these sound waves, an echo is sent back to the bat. Scientists believe bats send the sound out through their noses and mouths and can tell the size of the insect and its location by the echo.

Some animals sleep long and soundly; others take several brief naps. Each forest, meadow, or riverbank is home to two sets of animals. Some, like squirrels, and cows, and most birds, are active in broad daylight. Others like owls, rabbits, hedgehogs, weasels, and foxes prefer the half-light or darkness for finding their food. Some animals have become nocturnal as a way of protecting themselves from predators, while others use the cover of darkness to hunt their prey.

Bats are not blind, but they rely more on their keen sense of hearing to fly at night and hunt insects.

What happens to animals when the temperature turns cold?

As winter approaches, the temperature drops. The last flowers fade and the leaves fall from the trees. Soon frost and snow will cover the ground in many areas. What will the animals do? What will they eat? Each species has its own way of coping with winter.

Foxes have to travel farther during the winter because it's harder to find food.

Birds that stay in the north puff up their feathers to keep a layer of warm air between their skin and the cold air outside. Feathers provide excellent insulation against the cold. That's why coats and quilts are often filled with feathers. Many birds who can't survive the cold migrate to warmer climates.

As winter approaches, small mammals start to eat constantly to put on as much fat as they can. A thick layer of fat will help keep them alive when food becomes scarce. Their winter coats will help insulate them from the cold. Some animals work hard during the summer and fall storing food for the winter. Field mice hide seeds in their nests, and moles keep worms in their burrows. Squirrels hoard food in many places. These animals remain active nearly all winter, sleeping extra long only when it's very cold. Just like birds, they have to eat a lot to stay alive.

Magpies feed on insects and seeds. They will also eat the carcasses of dead animals.

In autumn, animals eat their fill.

In autumn, small mammals eat acorns, chokecherries, huckleberries, and other fruits and nuts.

Some of the animals in these pictures sleep right through the winter, after eating all the food they can. During hibernation, the body temperature of animals drops and they live off of stored fat.

Squirrels hide food that they dig up and eat during the winter.

The dormouse of Europe and Asia eats seeds and berries for six months and hibernates for six months. Bears eat anything they can find to prepare for the long winter ahead—berries, insects, leaves, twigs, and small mammals. Thousands of bats may spend the winter together in the shelter of a cave.

Animals struggle to stay alive in the cold.

Snails seal the entrances of their shells with a layer of mucus.

How can fish live below the ice? As long as the water stays above freezing, they can survive but the cold slows many of them down. The cold-sensitive tench, a fish from Europe, sinks into the mud as protection against the cold. But other fish manage to go about life as normal, while above them the world is in the grip of winter.

Many ducks will remain in cold climates as long as there are open bodies of water. They must dive into the cold water to find their food, but their water-proof feathers keep them warm. If the lakes and rivers freeze over completely, the ducks will be forced to fly farther south in search of open water.

Fish and other animals can survive beneath the frozen surface.

Many animals sleep the winter away.

Where are the insects in winter? Many of them die at the end of the summer after laying eggs that will survive the cold and hatch in the spring. Others spend the winter as a pupa, to emerge as an adult butterfly or moth in spring.

Some insects hide away. Beetle larvae burrow into the soil, and ants stay deep in their anthills. Ladybugs huddle in groups under dead leaves and in the cracks of window frames.

Snakes spend the winter together under stones or in caves.

Worker honey bees will live only about six weeks during the summer, but those who are alive when winter arrives will live through four months of hibernation in their hive.

Snakes, frogs, and snails must also find a safe place to spend the winter. Some snakes who had spent the summer alone now search out others of their kind and den up for the winter. Hundreds of snakes might pass the winter together in one warm den.

A tortoise buries itself in a deep hole for the winter.

Bees stay in their honey-filled hives.

Frogs sleep burrowed in the mud at the bottom of ponds, and snails find cracks in walls and under bricks to hide in. All these animals must slow down their heart and breathing rates and reduce their body temperatures, which in turn will reduce their need for food. They will hardly move until spring.

Insects find shelter in winter.

Inside winter shelters

When the days grow short, families of fat marmots go into their burrows. They block the entrances with earth and stones to keep out the cold. Then they fall into a deep sleep. Even when the temperature is as low as 15 degrees Fahrenheit (-10°C) outside, inside the burrow it's a cozy 45 degrees Fahrenheit (7°C). Here they will wait for the warmth of spring to arrive.

Bats hibernate in caves and attics. Their body temperature is as low as 40 degrees Fahrenheit (4.5°C).

The marmot's body slows down between October and March. It eats nothing and rolls up into a tight ball to stop body heat from escaping. Its temperature drops from 97 degrees Fahrenheit (36°C) to almost 40 degrees Fahrenheit (4.5°C). Its heart beats slowly, dropping from 100 beats to 4 beats per minute. It breathes only about once every six minutes to conserve the oxygen supply inside the burrow. If they get extremely cold, marmots wake up. If they didn't wake and move around a little to warm up, they could freeze to death in their burrows.

Although these animals don't hibernate, they do need shelter from cold temperatures. The deer mouse finds shelter in the walls. Squirrels curl up in nests they've built. Porcupines crawl into holes in rock piles.

Species hibernate for different lengths of time. The severity of the winter determines the length of time an animal will hibernate. If autumn is mild, animals will go into hibernation a little late, and if spring arrives early, they wake up sooner.

People should never disturb hibernating animals. Animals are in a very vulnerable state while hibernating and could easily be hurt by or die from exposure to the cold temperatures.

The warmth of spring wakes the sleepers.

Kangaroo rat

Each species has its own internal "alarm clock" that's set when hibernation begins. This clock carefully monitors the changes in the animal's body until it's time to wake up again.

What do bears do through the winter? They spend autumn eating, putting on a thick layer of fat before settling down in their dens. Bears don't actually hibernate, and can be awakened easily from their sleep. Their body temperature only drops a few degrees below normal.

Bear cubs are born during the winter. Newborn cubs are very small and unable to move far. When born, the cubs are only about the size of rats. During the winter they must eat and grow strong. When spring comes, mother and cubs will leave the den together.

Many animals slow down in the cold. Like bears, badgers retreat into their burrows during the coldest parts of winter, but don't really hibernate. White-tailed prairie dogs stay in their burrows from October to March. They awaken from time to time to feed on roots and seeds.

When spring arrives, both sleepers and hibernators wake up. Sunlight warms the earth, telling the sleepers it's time to get up. The bears, bats, and all the other hibernating and sleeping animals wake from their deep sleep when the temperature rises outside. Their hearts begin to beat faster and they breathe more deeply and more often. Their bodies begin to warm up. They've lost weight and must leave their shelters to search for food.

Groundhogs

Grizzly bears

Birds migrate to escape the winter.

Birds gather before migrating.

Many birds disappear at the beginning of autumn. They start on a long flight to warmer climates. This is called migration.

When the weather gets cold and snow begins to fall, food becomes more difficult to find. Life gets hard. Ducks and geese start to migrate from northern Canada and Alaska by mid-August, but farther south, birds wait until later. By autumn, swallows and blackbirds are gathering in flocks, ready to leave. Swallows and blackbirds fly in huge flocks, while ducks and geese fly in V formations. Flying in a V formation makes flight easier because none of the birds will have to fly in the air disturbed by the bird in front of it.

They fly to warmer climates.

Birds fly at different altitudes. Some small birds barely skim across the surface of the ground or the sea, whereas many birds may fly as high as 6,000 feet (1,830 m) above sea level. Geese can fly hundreds of miles a day, while hummingbirds may take many days to cover that distance, dodging birds of prey.

How do birds find their way? Some remember their flight paths from previous journeys. They always follow the same river valleys and coastlines, even though there may be shorter routes available. However, they are not all navigational experts, and sadly, some lose their way.

New Zealand shearwaters raise their young in New Zealand and then migrate as far north as British Columbia, Canada.

The golden plover leaves its home in Canada or Alaska and flies a long distance to South America.

Black swifts make short, summer migrations, flying hundreds of miles to get away from storms. They return to their nests days later, after the storm has passed.

The presence of birds can indicate the changing of seasons. After a long winter, seeing a robin means that spring is near. Ducks gathering together in late summer foretells the coming of winter.

Swallows gather on power lines before setting off on their journey south.

Birds find their way across vast distances.

Ducks rest by lakes and rivers along their migratory route.

Birds have different ways of navigating.
By day some birds follow the sun, and at night they follow the moon and stars, just as sailors sometimes do.

Many European birds go to Africa for the winter. They need all their strength to make the long journey.

Sailors use compasses and instruments called sextants to determine where they are, but birds calculate the routes they wish to take by instinct. Scientists think some birds navigate by using the Earth's magnetism. They can find their way as accurately as if they had a compass.

Sometimes scientists band birds to find out where they fly.

When they reach the Mediterranean, small birds take a rest before making the long flight across the sea. They'll stop again on the other side before flying across the Sahara Desert.

Studying how birds migrate can help scientists understand how to protect them.
Scientists who study birds are called ornithologists. Ornithologists note the names, numbers, and flight directions of different types of birds. Sometimes they put bands on birds' feet to find out their flight paths. This information can be valuable in understanding birds' behavior. Only by knowing what birds need to survive can an ornithologist determine the best way to help them.

Birds use air currents to fly higher and faster while conserving their own energy.

1. Black-headed gull 2. Skylark 3. Starling 4. Blackcap 5. Nightingale
6. Cuckoo 7. Hoopoe 8. Barn swallow 9. Crane 10. Oriole
11. Flycatcher 12. Black kite 13. White stork 14. Tern

Other winged travelers also migrate.

Birds aren't the only animals that migrate by air, although they are among the largest group. Bats, dragonflies, fragile butterflies, and even small spiders have been found at great altitudes, blowing in the air from one place to another.

Not all species of bats hibernate; some migrate to warmer climates. On autumn evenings, some bats begin their long journeys south. Female bats migrate with their young to countries like Mexico, where there are many caves and a favorable climate in which to spend the winter. Some bats are known to have migrated as far as 1,500 miles (2,400 km) south.

Butterflies are generally solitary insects that stay in one habitat. However, some butterflies, such as the monarch, migrate in large swarms to the mountains of Mexico. Many of the butterflies that begin migration do not complete the journey, dying along the way. But this does not stop other butterflies from embarking on the same grueling journey year after year.

Some delicate butterflies migrate.

The winged travelers eventually return. Most migratory animals return as soon as spring arrives. Once they have returned it is time to start all over, building nests and raising their young.

Monarch butterflies gather by the thousands and migrate from Canada and the United States to Mexico.

Marine animals are also on the move.

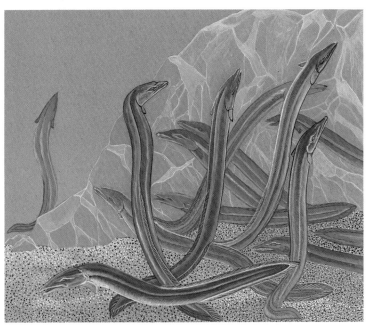
Eels

Sea turtles

Many animals travel for breeding purposes. Eels leave the rivers of Europe and migrate to the seaweed-rich Sargasso Sea near the Gulf of Mexico. There they lay their eggs and then die. The young eels have to undertake the long and dangerous journey back to Europe, where the cycle begins again.

Salmon spend their adult lives at sea and return to the river of their birth to spawn. After spawning, the adult salmon die. The young fish then make their way to the ocean. Green sea turtles swim 1,000 miles (1,600 km) through open ocean to find a sandy beach on which to lay their eggs.

Salmon and crayfish

Four-legged animals travel to new pastures.

It's far more tiring to travel overland than to swim or fly. This may be why only a few land animals migrate. On a small scale, frogs and toads must leave the undergrowth and move to watery places if they are to mate and lay eggs. The tadpoles that hatch will eventually grow legs and migrate onto land to start the cycle all over again.

Caribou

Migrating wildebeests

Some mammals are forced to migrate to find food. Herds of African wildebeest migrate together. They move when the grass becomes too scarce and the water holes dry up. They go to wetter areas to find fresh grass and plentiful water. Many are eaten by lions, and some are captured by crocodiles when they cross rivers.

In cold places like Siberia, Greenland, and Alaska, caribou, which are called reindeer in Europe, travel hundreds of miles south to find food. If they didn't undertake this journey, they would starve. The long winters and frozen ground of the north make food scarce. Only the strongest animals can survive this journey; the weak will be hunted by wolves.

In winter, ibexes and chamois in Europe leave their craggy mountain homes to live in the forests of lower elevations. There they can find food and are protected against the worst of the weather. The forest trees shield animals from the wind and snow, while their thick winter coats keep them warm.

The animal calendar

All animals are affected by the seasons. Changes in the length of the days, in temperatures, and even in the moon's cycles send signals to an animal's brain. When the temperature begins to fall or days shorten, animals prepare to hibernate or migrate. When the days lengthen in spring, birds start building their nests and many animals give birth. This will allow their young to take advantage of the abundant food of summer.

The color of an ermine's coat changes seasonally. This keeps it from being seen by its prey.

Springtime is the best time of year for many animals to give birth. At this time of the year, there is usually plenty of food available for parents to feed their young. The milder weather gives the babies a better chance for survival.

Animals give birth in different places. Woodchucks, prairie dogs, and rabbits produce their young underground in cozy nests. Larger animals, like deer and wild boar, find sheltered places in the forest.

Summer and autumn are times for learning. These are the seasons when young animals must learn the skills they'll need to survive without their parents. By the time winter arrives, they must be strong enough and clever enough to cope with cold and hunger. Only the fittest will survive to have families themselves.

When snow begins to fall, it becomes more difficult for these animals to find food.

Its small size doesn't stop the weasel from attacking animals larger than itself.

The balance of nature

Everything has its place in nature. Whether it lives in the mountains, the rivers, or the oceans, each plant and animal has an important role to play in its environment. Plants make their food from water, minerals in the soil, and carbon dioxide in the air. Similar to most living things, they also need energy from the sun. In turn, plants become food for many animals and produce oxygen for animals to breathe.

The delicate balance of nature must be maintained. Each group of animals must be kept in check so that no one species becomes too numerous and overwhelms others. This is the importance of the food chain. Herbivores eat plants, and carnivores eat herbivores. Each species plays its part in the cycle of life. Some plants and animals feed on dead things. As dead things decompose, nutrients return to the soil to enrich it, which in turn helps plants grow. Insects help pollinate plants so they will produce fruit and seeds. Even the humble earthworm has an important task. It tills the soil to keep it healthy and well-drained.

A single oak tree provides food and shelter for many animals. They live in nests among the branches, in holes in the trunk, or in burrows among the roots. Some live side by side; others compete for food. Wasps chew the wood to make pulp to build their papery nests. Woodpeckers hammer out their homes in the trunk and eat insects that might harm the tree. Plants and animals depend on each other to make a rich community and keep a healthy balance of all living things.

Wildlife Alert!

Animals have inhabited the Earth for millions of years. In each generation, the strongest and fittest have survived to mate and have young of their own. The offspring then had to work hard to survive themselves. Only a small number of all the species that have ever lived are still alive today.

Komodo dragons are no longer hunted today, but many of the animals they eat are. If it is to survive, the dragon's food source must be protected.

Now human activities are poisoning the land, the rivers, and the air. Because of this, the homes of many animals are no longer safe for them, and a great number of different species have become, or are in danger of becoming, extinct.

Dodo birds became extinct when sailors landed on the island of Mauritius and hunted them for food.

Humans are a threat to gavials because they're killed for their skins.

Species are massacred.

During the 19th century, millions of bison were killed on the Great Plains by hunters who sold the skins and left the carcasses to rot. Today, the bison populations are protected in national parks or in private reserves and are making a comeback.

Some people still hunt endangered animals for pleasure or profit. Crocodiles and leopards are hunted for their skins, while elephants are hunted for their ivory tusks and rhinoceroses for their horns.

Baby harp seals are hunted for their white fur.

The blue whale has been hunted for thousands of years by Inuit people who throw harpoons from open boats. Few people need to eat whales today, yet they are still hunted. The development of harpoons with exploding heads and factory ships to process the whales has led to all large whales being endangered.

Bison

Blue whale

Endangered animals in the forest

Some animals lose their shelter and their source of food when forests are cleared to grow crops or to provide grazing for domestic animals. Without tree cover, the soil may become dry and dusty. It may eventually blow away, leaving a barren wasteland.

Pandas are among the rarest mammals in the world. They live only in the bamboo forests of China. The destruction of these forests are threatening the pandas' survival. Chinese naturalists now rescue the pandas when they can.

The black-footed ferret is the rarest mammal in North America. Fewer prairie dogs, the ferret's favorite food, has threatened its survival. Scientists believed it was extinct until they discovered it in Wyoming.

Animals are at risk from many human activities.

Only a few gorillas remain in tropical Africa. Many have been killed as trophies, and young gorillas have been sold to zoos. Their numbers are slow to increase because female gorillas have babies only every four or five years and because of the destruction of the forests where they live.

The bald eagle, the symbol of the United States, was nearly extinct. Its numbers had fallen mainly because the fish on which it feeds had been contaminated by pesticides like DDT. The eagle was then given protected status and most uses of DDT were banned. Since then, the bald eagle has made a comeback.

Sadly, other birds, like the California condor, haven't had such luck. There are fewer than 50 of these birds left.

River and sea pollution are serious threats.

The population of seals, sea otters, turtles, and manatees has decreased greatly over the years. Human activities like oil exploration and dam building have made life extremely difficult for these and other aquatic animals.

Many turtles, like the leatherback turtles below, are becoming endangered. Pollution poisons the water where they live, and human activity destroys the sandy beaches where they must lay their eggs.

Manatees, aquatic mammals found along the warm coastal waters of West Africa, the Caribbean, and Florida, are slowly decreasing in number. They are often fatally injured by the propellers of boats that race through the waters where they live.

Some countries have declared some rivers and coastlines nature reserves in order to protect animals like sea otters. Otters were once hunted for their furs or killed simply because people believed that they caught and ate too many fish.

Protecting animals

Animals are protected in their natural habitats in a number of ways. Wildlife refuges, national parks, wilderness areas, and national forests are all specially created sanctuaries that give animals safe homes. Laws have been passed that make it a crime to hunt certain animals. Perhaps someday the rare lynx hunting in the snow will again be a common sight, and elephants and rhinoceroses will graze safely on the African plains.

Here are some things people can do to help. Avoid using tropical hardwoods like mahogany. Using them means destroying precious rain forests. Use recycled paper. Paper is made from wood, and millions of trees are cut down every year to make paper. The labels of most paper products will say whether or not recycled paper was used. Also, don't litter. Litter could harm a living creature. Humans share this planet with animals, and people can make it a better place.

EXPLORE AND PLAY

Intriguing facts, activities, games, a quiz, a glossary, and addresses of places to visit, followed by the index

■ **Did you know?**

Prehistoric people hunted with sharpened stones and used the bones of their prey to make tools.

The nautilus can float at different depths in the water by controlling the amount of air in the shell chambers behind its body.

The tenrec of Madagascar is a small insectivore that resembles the hedgehog. When it's upset, it shakes the prickly spines on its back to make a clicking sound.

Armadillos are covered with a protective armor of bony plates. The nine-banded armadillo is the only species in the entire United States.

■ **Animal identification**

Can you identify these animals? The answers are below.

1. This large dinosaur had a double row of bony plates on its back.

2. This animal is born with sharp quills on its back and is mostly active at night.

3. This animal has orange and black wings and flies to Mexico in the autumn.

4. This animal has tusks, flippers, and whiskers.

Answers:
1. Stegosaurus 2. Porcupine 3. Monarch butterfly 4. Walrus

■ **Did you know?**

The Colorado potato beetle feeds on tomato leaves as well as potato leaves. This beetle spends the winter buried in the soil and emerges in spring to lay its eggs.

How echidnas are born
Echidnas, or spiny anteaters, are egg-laying mammals called monotremes. When a female echidna is ready to have a baby, she grows a flap of skin on her abdomen. This skin will form a pouch. When the egg is laid, the echidna wriggles it into the pouch. After 10 days, the young echidna hatches from its egg and laps up milk from glands in the pouch. The young echidna will live for several weeks in the mother's pouch before venturing outside.

Echidnas and platypuses are the only egg-laying mammals. They both live in Australia and Tasmania. Echidnas are also found in New Guinea. Both animals are protected by law.

Did you know?

Caterpillars are very particular about what they eat. To ensure the caterpillar's survival, butterflies must lay their eggs not only on the right plant, but also on the right part of the plant.

The crab spider matches the colors of its surroundings. It lurks among flowers, looking like part of the plant. When a bee or butterfly settles beside it, the spider catches and kills its prey by pumping poison into its body.

Head or tail?
The false eyes on this frog's thighs confuse its enemies, giving it time to escape from harm.

Animal identification

Can you identify these animals? The answers are below.

1. This animal plucks its own chest hair to line the burrow for its babies.

2. This bird lays its eggs in other birds' nests.

3. This animal cuts down trees using teeth that grow all its life.

4. This animal howls to keep in touch with other members of its pack.

Answers: 1. Rabbit 2. Cuckoo 3. Beaver 4. Wolf

Did you know?

Reptiles change their skin from time to time in a process called molting. A snake may molt due to worn-out skin or because snake scales do not stretch when it grows larger. A sign that a snake is about to shed is that the spectacle scales covering the snake's eyes become milk colored. At this time, the snake is defenseless because it can't see. Shedding begins when the snake rubs against small rocks to loosen the dead skin. Once the skin has caught on something, the snake is able to slip out with its new scales looking clean and bright.

Some lizards can shed their tails and escape in an emergency, leaving their predators with a wriggling tail tip.

This insect may look dangerous, but it doesn't sting. It only looks like a wasp to frighten its enemies away.

■ **Did you know?**

The paradise fish builds its nest from saliva and bubbles that rise to the surface to form a raft. The eggs are laid and fertilized, then placed in the raft and sealed. The male looks after the eggs until they hatch into hundreds of tiny fish.

Long-distance migration
During its 20-year life span, a black swift may fly as many as 250,000 miles (400,000 km), a distance greater than that between the Earth and the moon.

1. Kallima butterfly
2. Hawk moth caterpillar
3. Peppered moth caterpillar
4. Purple emperor chrysalis
5. Silk moth caterpillar

Insect camouflage
Butterflies and caterpillars are the favorite foods of lizards, birds, snakes, and monkeys. Luckily, they have developed camouflage that gives them a chance for survival. They pretend to be thorns, leaves, animal droppings, and even other animals. How many butterflies, moths, and caterpillars can you find in the picture below?

■ **Did you know?**

Northern fur seals migrate large distances. They spend their summers in the Bering Sea and swim south 6,200 miles (10,000 km) to the warmer waters off California to spend the winters. When searching for food, the seals can dive as deep as 450 feet (135 m) and remain submerged for several minutes. Northern fur seals spend most of their lives in the water, returning to land only during the breeding season.

6. Female brimstone butterfly
7. Black hairstreak chrysalis
8. Lappets moth
9. Death's-head hawk moth
10. Poplar hawk moth caterpillar

The corncrake lives in wild, marsh grasses of Europe and Asia. The draining of the marshes and wetlands, which these birds need to survive, is pushing them toward extinction.

Destruction
Thousands of species continue to be threatened with extinction because humans destroy animal and plant habitats by building new houses and roadways.

■ **Did you know?**

The number of African rhinoceroses is decreasing rapidly. Even though they are a protected species, poachers hunt rhinoceroses for their horns, skin, and blood, which some people use in medicines. In an attempt to prevent their extinction, many countries have signed treaties that ban the sale of products made from rhinoceroses.

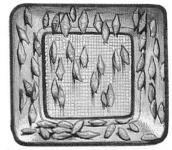

The beautiful coloring of many animals is a threat to their survival. Poachers trap and kill animals like cheetahs for their beautiful furs, which are then made into coats. Even something that seems harmless, like collecting butterflies, can cause great damage to that species and the rest of the food chain.

Humpback whale

Can elephants be saved from extinction?
No one knows the answer to that question. The population of African elephants dropped from 1.3 million in 1979 to 600,000 in 1989. Since countries around the world began to ban the sale of ivory in 1989, this rapid decline has slowed. Despite these bans, thousands of elephants are still killed every year for their ivory tusks. If this continues, there soon won't be any elephants left.

Quiz

The answers are at the bottom of the page.

1. Diplodocus was a giant
a. carnivore.
b. herbivore.
c. omnivore.

2. Pterosaurs were
a. birds.
b. reptiles.
c. mammals.

3. The first mammals appeared
a. after the Ice Age.
b. 220 million years ago.
c. in the sea.

4. Turtle eggs
a. are incubated by the mother.
b. float in the sea.
c. are buried in the sand.

5. Which of these animals feeds its young on milk?
a. crocodile
b. whale
c. shark

6. Which of these birds is nocturnal?
a. bald eagle
b. barn swallow
c. owl

7. Which of these young animals will never know its parents?
a. cod
b. wolf
c. swallow

8. Baby bears are reared by
a. their mothers.
b. both parents.
c. first the mother, and then the father.

9. When kangaroos are born
a. they can jump immediately.
b. they crawl into their mother's pouch.
c. they are protected and warmed by the group.

10. What does a squirrel do in winter?
a. hibernate in its nest
b. sleep and eat alternately
c. migrate to a warmer area

11. Which of these do not migrate in winter?
a. northern fur seal
b. grizzly bear
c. monarch butterfly

12. Which of these animals doesn't have a winter sleep?
a. porcupine
b. dormouse
c. fox

13. How long does a marmot usually hibernate?
a. three months
b. six months
c. four months

14. Bats sleep
a. upside down.
b. standing up.
c. lying down.

15. Insect-eating bats can find their way in the dark because of their
a. extremely good vision.
b. keen sense of smell.
c. ability to use echolocation.

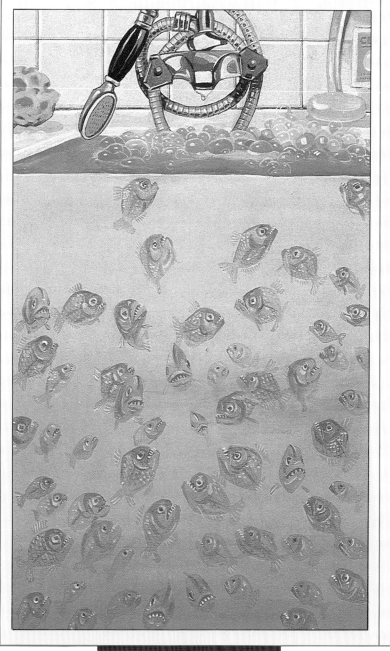

Fish can't really swim up a bathtub drain.

Answers: 1b, 2b, 3b, 4c, 5b, 6c, 7a, 8a, 9b, 10b, 11b, 12c, 13b, 14a, 15c

In a manner of speaking

Humans often compare themselves and their friends to animals, contrasting their qualities, faults, and behavior to that of other creatures. Here are a few common examples that you may have heard.

An old fox is a crafty person.
A bird brain is someone with little common sense.
A shark describes a merciless person.
As prickly as a porcupine describes an unfriendly person.
A chameleon is someone who changes very quickly.

Spotted owl

A night owl is a person who stays up late at night.
A parrot is someone who repeats what other people say, often without understanding it or thinking about it.
A cold fish is someone who doesn't show much feeling.
A snake is an untrustworthy person.
A hyena is someone who attacks weak and

Are you . . .

Blind as a bat? Someone with poor eyesight
Wise as an owl? Full of intelligence
Hungry as a wolf? Extremely hungry
Busy as a bee? Buzzing around all day
Sly as a fox? Full of tricks
Happy as a lark? In a good mood
Slippery as an eel? Hard to catch
Proud as a peacock? Enjoy showing off

Do you . . .

Have ants in your pants? Can't sit still
Have the memory of an elephant? Never forget anything

Tarantula

People use animal expressions to describe how they move.

In a beeline means to move straight ahead quickly.
As the crow flies means to take the shortest, most direct route.
At a snail's pace means to move extremely slow.

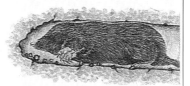

Can you . . .

Climb like a monkey? With speed and agility
Run like a rabbit? Very quickly
Eat like a horse? Eat a lot

Do you . . .

Have eyes like a hawk? Have sharp vision
Run with the pack? Join in with others, sometimes doing silly or bad things

A leopard can't change its spots. People are what they are, and no one can expect them to change their ways overnight.
A wolf in sheep's clothing is someone who's pretending to be kind and gentle but really has bad motives.
To take to something like a duck to water means to get used to it right away.

■ If you're interested in nature, there are many things you can do to learn more about the plants and animals around you.

You could join a wildlife group, like the National Audubon Society.

To closely observe nature, take a nature walk. Here are some tips.

Walk into the wind so that animals will be less likely to smell you and run away.

Use the cover of trees and walk with the sun behind you.

What special equipment and clothing will you need?

Choose waterproof shoes or boots and a warm, waterproof jacket. Take a note pad, pencils, and maybe a magnifying glass. A mayonnaise jar works well as a place to observe small animals, but remember to release all animals after you have looked at them. Take a whistle. It's a good way of drawing attention to yourself if you get lost, become hurt, or have trouble.

A flashlight might be necessary if nightfall is approaching. You may not need all the equipment on this jacket, but it is important to always be prepared when going into the wilderness.

Plan an outing with a friend, but remember to always tell an adult where you're going and what time you expect to be back. Better still, ask adults to go along. They will learn as much as you. There are always new things to enjoy and discover in nature.

The size of the entrance to a burrow indicates if it belongs to a badger or a much smaller prairie dog.

Hiking equipment

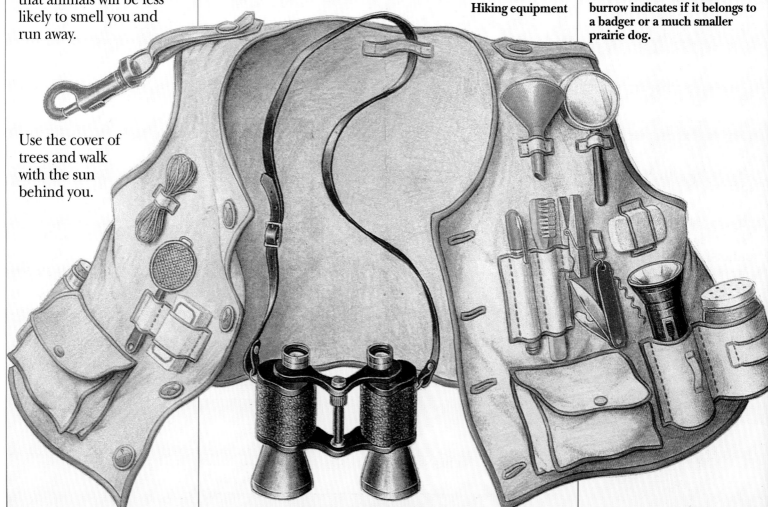

You may decide to look for tracks and signs on your nature walk.
Look for nuts and acorns. You can often see a squirrel's teeth marks on the shells. Look at the bark of young trees to see if they've been chewed by rabbits or deer. You may also see animal tracks in muddy ground.

If you wait quietly, a number of animals may pass by. Mice may pass within inches of your feet. If you're near a stream, you may see a field mouse or a muskrat. If you're really lucky, you may even glimpse a bald eagle.

You can also invite nature to your yard. In winter, you can leave food out for the birds. Corn and nuts left out will also attract squirrels. Salt licks will attract deer in wooded areas. In summer, plant flowers to attract butterflies and hummingbirds.

Look at this zoologist's notebook. Can you see how the drawing is created?

Practice makes perfect.
Draw birds in your yard or in a park. Try to capture the way they move and take notes on the way they behave.

Below are the tracks made by many common animals. Look for them when you're out on a walk.

What things can you collect? Always make sure that you aren't breaking any laws when you start a collection. Talk to a park ranger or naturalist to find out the laws. Many plants and animals are protected.

You can take home souvenirs from a walk. If you're careful with delicate treasures, such as fallen leaves and flower petals, you can press them in a book and preserve them into the future.

Molted flight feathers, skulls, and animal bones are lucky finds. But be careful; it is illegal to collect the feathers of eagles and some other birds.

Enjoy the beauty of flowers, but don't pick them. Leave them for other people to enjoy.

■ Glossary

Algae: plants that live in water, from tiny one-celled plants to giant seaweed.

Bacteria: tiny one-celled organisms. Some live in animals' bodies and do useful work; some cause disease.

Baleen: the sieve-like plates that trap food in the mouths of some whales.

Breed: to reproduce offspring.

Camouflage: the colors or pattern of an animal's skin or fur that helps it blend into its environment.

Carbon dioxide: a gas in the air that plants need to make their food. Animals breathe it out.

Carcass: the dead body of any animal, especially one that was killed for food.

Cold-blooded: describes any animal that relies on heat from the sun to regulate its body temperature.

Elephant seal

Contract: to narrow. When it's light outside, some animals' pupils contract to keep too much light from getting in. Pupils can also dilate, or widen, to let in more light.

Courtship: the process animals go through to attract mates.

Display: to show off, usually done by animals to impress or frighten.

Doe: a female deer, rabbit, or hare.

Echolocation: a system used by some animals for determining their location by emitting sounds and listening for the returning echo.

Egg: the female sex cell. The egg is fertilized by the sperm.

Endangered: term used to describe a species in danger of extinction.

Environment: the soil, water, air, and everything else that makes up an organism's surroundings.

Evolution: the gradual change in the characteristics of a species over generations.

Extinct: no longer existing. Today some species of animals face extinction because of overhunting and the destruction of their habitat.

Fertilization: when the male and female sex cells join to make a new life.

Fledge: to feed and care for a young bird until it's ready to fly. Also, to grow flight feathers.

Fossil: the remains of a plant or animal that have turned to stone over millions of years.

Fry: a newly hatched fish.

Habitat: the place, or environment, that an animal needs to survive.

Incisor teeth: the front, biting teeth that work like chisels in rodents and other mammals.

Instinct: the knowledge that animals are born with which helps them survive.

Kit: the cub of various small mammals such as beavers, ferrets, and foxes.

Larva: the eating stage in the life cycle of a butterfly or moth.

Marsupial: mammals whose young are born undeveloped and spend the first weeks of life in their mother's pouch.

Mate: to come together in order to produce young.

Metamorphosis: the complete change of shape that happens to some animals before they become adults.

Migration: the movement of animals to find food or warmer climates. Migration can take place daily or yearly.

Nutrient: any part of a food that provides nourishment.

Organism: a general term used for any plant or animal.

Ornithologist: a person who studies birds.

Oxygen: a gas in the air that humans and animals need to breathe.

Paleontologist: someone who studies fossils.
Placenta: an organ formed inside most mammals during pregnancy that provides food for the young before birth.
Predator: any animal that catches and eats other animals.
Prey: any animal that is hunted and eaten by another animal.
Pride: a group of lions.
Pupa: the stage of an insect's life during which it goes through metamorphosis and becomes an adult.

Rodent: mammals whose incisor teeth grow throughout their lives.

Spawn: term, usually referring to fish, used to describe the process of egg fertilization.

Species: a group of related organisms whose members can breed with each other.

Sperm: the male sex cell. The sperm fertilizes the egg.

Warm-blooded: describes an animal that can internally regulate its body temperature. Its body is usually warmer than its surroundings.

Zoologist: someone who studies animals. The word zoo, short for zoological garden, comes from the same root word.

Gorillas live in troops, cattle live in herds, and wolves live in packs.

Here is a list of organizations to join and museums and zoos to visit to learn more about animals. Ask your local library for more information about places to visit in your area.

Smithsonian Institution
1000 Jefferson Drive SW
Washington, D.C. 20560

The San Diego Wild Animal Park
15500 San Pasqual Valley Road
Escondido, CA 92027-7017

The Academy of Natural Sciences
1900 Benjamin Franklin Parkway
Philadelphia, PA 19103

University of Oregon Museum of
Natural History
1680 East 15th Avenue
1224 University of Oregon
Eugene, OR 97403-1224

Canadian Museum of Nature
P.O. Box 3443, Station D
Ottawa, Ontario
Canada K1P 6P4

Sea World Florida
7007 Sea World Drive
Orlando, FL 32821-8097

The World Center for Birds of Prey
Velma Morrison Interpretive Center
566 West Flying Hawk Lane
Boise, ID 83709

The Denver Zoo
2300 Steele Street
Denver, CO 80205

African Wildlife Foundation
1400 Sixteenth Street, NW Suite 120
Washington, D.C. 20036

The Nature Conservancy
4245 North Fairfax Drive, Suite 100
Arlington, VA 22203-1606

National Wildlife Federation
8925 Leesburg Pike
Vienna, VA 22184

National Audubon Society
700 Broadway
New York, NY 10003

INDEX

The entries in **bold** refer to whole chapters on the subject.

A
Algae, 4, 31, 70
Allosaurus, 8, 9
Amphibians, 5, 13
Ankylosaurid, 9
Ant, 16, 37, 43, 67
Anteater, 12, 27
Antelope, 31
Ape, 36
Arctic fox, 29
Autumn, 40, 41, 44, 45, 46, 49, 52, 62

B
Babies, Changing with the times, 12, Animal families can be large or small, 13, Not all parents stay with their young, 14, Mammals and birds are caring parents, 15; *see also* 6, 16, 18, 21, 23, 28, 34, 39, 45, 50, 57, 62, 64
Bacteria, 4, 70
Badger, 20, 28, 35, 38, 45, 68, 69
Bald eagle, 57, 69
Baleen, 27, 70
Bat, 24, 39, 41, 44, 45, 49, 67
Bear, 14, 16, 38, 41, 45
Beaver, 12, 22–23
Bee, 16, 35, 43, 63, 67
Birds, 13, 14, 15, 16, 18, 19, 27, 30, 31, 34, 40, 46–48, 52, 64, 69
Bison, 55
Blackbird, 18, 46
Boa constrictor, 26
Boar, 26, 52
Bowerbird, 18, 35
Brachiosaurus, 7
Brontosaurus, 8
Butterfly, 13, 24, 28, 43, 49, 63, 64, 65, 69

C
California condor, 57
Camouflage, Many animals can disguise themselves, 28, Some animals match their backgrounds, 29, Animals' spots and stripes hide their shapes, 30, All disguises help animals survive, 31, Trickery on the sea bed, 32, Seeing isn't always believing, 33; *see also* 17, 63, 64, 70
Canine, 24, 26
Carbon dioxide, 53, 70
Carcass, 24, 70
Caribou, 51
Carnivore, 9, 24, 26, 53
Caterpillar, 13, 63, 64
Chameleon, 27, 29, 67
Chamois, 51
Cheetah, 28, 31, 65
Chimpanzee, 14, 17, 36
Cod, 14
Cold-blooded, 5, 9, 70
Communication, Colors are signals for mating and warning of danger, 34–35, Animals communicate in many different ways, 36, Some animals use scent and sound, 37
Compsognathus, 7
Coot, 19
Corncrake, 65
Courtship, 34, 70
Crab, 27, 32
Crocodile, 8, 9, 12, 14, 51, 55
Cuckoo, 18, 48

D
Decompose, 11, 53
Deer, 25, 52, 69
Digestion, 25, 26
Dimetrodon, 5
Dinosaurs, 7–10, 12
Diplodocus, 7
Dodo, 54
Doe, 21, 70
Dogfish, 13
Dormouse, 41
Duck, 42, 46, 47, 48, 67

E
Eagles, 19, 27, 57, 69
Earthworm, 53
Echidna, 62
Echolocation, 39, 70
Eel, 50, 67
Elephant, 11, 12, 15, 25, 26, 55, 59, 65, 67
Endangered animals, Wildlife Alert, 54, Species are massacred, 55, Endangered animals in the forest, 56, Animals are at risk from many human activities, 57, River and sea pollution are serious threats, 58, Protecting animals, 59
Evolution, 5, 70
Extinction, 4, 5, 10, 11, 12, 54, 57, 65, 70
Eyes, 38–39, 67

F
False eyes, 33, 63
Fennec fox, 21, 31
Fertilization, 12, 13, 70
Fish, 25, 33, 35, 57, 58, 67
Fledge, 18, 70
Food chain, 53
Fossil, 4, 5, 6, 8, 10, 11, 70
Frog, 10, 13, 16, 27, 28, 31, 39, 43, 51, 63
Fry, 13, 70

G
Geese, 46, 47
Glowworm, 35
Glyptodont, 10
Golden plover, 47
Gorilla, 14, 36, 57
Grooming, 36
Guinea pig, 25

H
Hamster, 25
Hedgehog, 24, 37, 39, 69
Herbivore, 24, 25, 53
Hibernation, In autumn, animals eat their fill, 41, Animals struggle to stay alive in the cold, 42, Many animals sleep the winter away, 43, The warmth of spring wakes the sleepers, 45, Inside winter shelters, 46
Horse, 25, 67

I
Ibex, 51
Ichthyosaur, 6
Incisor teeth, 22, 25, 70
Insect, 13, 27, 30, 39, 43, 49, 53, 63, 64
Insectivore, 24, 62
Instinct, 15, 48, 70
Insulate, 9, 40
Ivory, 26, 55, 65

J
Jerboa, 30, 31

K
Kangaroo, 12, 13
Kit, 23, 70
Komodo dragon, 54

L
Ladybug, 43
Larva, 13, 16, 43, 70
Lemur, 35
Leopard, 30, 55, 67
Lion, 12, 14, 15, 51
Lizard, 9, 31, 63, 64
Lynx, 59

M
Magpie, 40
Maiasaurus, 8
Mammals, 9, 10, 11, 12, 13, 14, 15, 20–21, 31, 39, 40, 41, 51, 56, 58, 62
Mammoth, 11

Manatee, 58
Mandrill, 34
Marmot, 16, 44
Marsupials, 13, 71
Mating, 12, 14, 21, 34, 35, 54, 71
Metamorphosis, 13, 71
Migration, Birds migrate to escape the winter, 46, They fly to warmer climates, 47, Birds find their way across vast distances, 48, Other winged travelers also migrate, 49, Marine animals are also on the move, 50, Four-legged animals travel to new pastures, 51; *see also* 52, 64, 71
Molar, 24, 25
Mole, 21, 24, 37, 40
Molting, 63
Monarch butterfly, 49
Monkey, 12, 36, 64, 67
Monotreme, 62
Mouse, 38, 44, 67, 69
Mussels, 16

N
Nature, The animal calendar, 52, The balance of nature, 53; *see also* 68–69
Nautilus, 4, 62
Nests, 16, 17, 18–19, 47, 52, 53
Nocturnal, Nighttime adventurers have senses that can penetrate the dark, 38–39; *see also* 21, 31, 35
Nutrients, 53, 71

O
Octopus, 33
Omnivore, 24
Organism, 71
Ornithologist, 48, 71
Ornithopods, 8
Ostrich, 15
Otter, 17, 58
Owl, 29, 38, 67, 69
Oxygen, 4, 5, 9, 44, 53, 71
Oyster, 16

P
Paleontologist, 10, 71
Panda, 56
Paradise fish, 64
Peacock, 34, 67
Placenta, 13, 71
Platypus, 21, 62
Plesiosaur, 6
Pollution, 58
Porcupine, 25, 38, 44, 67
Predator, 10, 26, 28, 30, 31, 35, 71
Prehistoric animals, The beginning of life on Earth, 4, The first plants and animals lived in the sea, 5, The age of giants, 6, The amazing dinosaurs, 7, They ruled the world for 160 million years and then disappeared, 8–9, Early mammals, 10, Some extinct mammals have relatives living today, 11
Prey, 8, 19, 24, 25, 26, 27, 30, 31, 39, 71
Pride, 14, 15, 71
Proceratops, 8
Pterosaur, 6
Pupa, 43, 71
Python, 26

R
Rabbit, 16, 21, 24, 37, 39, 52, 67, 69
Raccoon, 25
Radula, 27
Rat, 25, 69
Rattlesnake, 26
Ray, 13, 25
Razor fish, 33
Reindeer, 51

Reptile, 5, 6, 7, 9–10, 13, 63
Rhinoceros, 9, 10, 55, 59, 65
Rodent, 25, 71

S
Saber-toothed tiger, 10
Salamander, 13
Saliva, 16, 64
Salmon, 50
Scent, 36, 37
Scorpion, 31
Sea, 4–5, 32–33, 35, 58
Seagull, 34
Seahorse, 27
Seal, 55, 58, 64
Sea urchin, 25
Shark, 24, 25, 67
Shelter, Shelter from the outside world, 16, Homes can be many shapes and sizes, 17, Birds are master builders, 18, There are many ways to build a nest, 19, Some mammals make burrows that they line with grass and moss, 20–21, The beaver is a skilled architect, 22, An underwater entrance leads to a dry nest, 23
Shrew, 21, 24, 37
Skink, 13
Sloth, 31
Snail, 4, 16, 27, 32, 42, 43, 67
Snake, 9, 13, 16, 26, 30, 31, 39, 43, 63, 64, 67
Snowy owl, 29
Spawn, 50, 71
Species, 4, 12, 14, 45, 53, 54, 71
Sperm, 12, 13, 71
Spider, 17, 49, 63
Squirrel, 16, 17, 39, 40, 41, 44, 69
Stegosaurus, 8, 9
Stickleback, 17
Stonefish, 33
Swallow, 19, 46, 47, 48
Swift, 18, 47, 64

T
Tadpole, 13, 14, 51
Tailorbird, 19
Teeth, Teeth are used for killing and eating, 24, Animals have different kinds of teeth, 25, Tusks and fangs are teeth too, 26, Some animals have no teeth, 27
Tenrec, 62
Territory, 22, 36, 37
Tiger, 10, 15, 24, 30
Toad, 14, 16, 27, 39, 51
Tongue, 25, 27
Tortoise, 9, 43
Tracks, 69
Triceratops, 9
Trilobites, 4, 5
Turtle, 8, 50, 58
Tusks, 11, 26, 55, 65

V
Vinatherium, 10

W
Walrus, 26
Warm-blooded, 15, 71
Wasp, 35, 53
Weasel, 21, 29, 39, 52, 69
Weaverbird, 19
Whale, 13, 27, 55
Wildebeest, 51
Wolf, 36, 51, 67
Woodpecker, 19, 27, 53

Z
Zebra, 30
Zoologist, 69, 71